SLITHERING SNAKES

BOA CONSTRICTOR

READY TO SQUEEZE

BY NATALIE HUMPHREY

Enslow PUBLISHING

DISCOVER!

Please visit our website, www.enslow.com. For a free color catalog of all our high-quality books, call toll free 1-800-398-2504 or fax 1-877-980-4454.

Cataloging-in-Publication Data

Names: Humphrey, Natalie.
Title: Boa constrictor: ready to squeeze / Natalie Humphrey.
Description: New York : Enslow Publishing, 2021. | Series: Slithering snakes | Includes glossary and index.
Identifiers: ISBN 9781978517714 (pbk.) | ISBN 9781978517738 (library bound) | ISBN 9781978517721 (6 pack)
Subjects: LCSH: Boa constrictor—Juvenile literature. | Snakes—Juvenile literature.
Classification: LCC QL666.O63 H85 2021 | DDC 597.96'7—dc23

Published in 2021 by
Enslow Publishing
101 West 23rd Street, Suite #240
New York, NY 10011

Designer: Sarah Liddell
Editor: Natalie Humphrey

Photo credits: Cover, p. 1 (boa constrictor) Nicole Hollenstein/Shutterstock.com; background pattern used throughout Ksusha Dusmikeeva/Shutterstock.com; background texture used throughout Lukasz Szwaj/Shutterstock.com; p. 5 GlobalP/iStock/Getty Images Plus/Getty Images; p. 7 JH Pete Carmichael/The Image Bank/Getty Images Plus/Getty Images; p. 9 cynoclub/iStock/Getty Images Plus/Getty Images; p. 11 jamcgraw/iStock/Getty Images Plus/Getty Images; p. 13 fivespots/Shutterstock.com; p. 15 Vitor Marigo/Aurora Photos/Getty Images; p. 17 Michael Weber/Getty Images; p. 19 Matt Jeppson/Shutterstock.com; p. 21 Marko Tanasijevic/Shutterstock.com.

Portions of this work were originally authored by Cede Jones and published as *Boa Constrictor*. All new material this edition authored by Natalie Humphrey.

Printed in the United States of America

CONTENTS

Boldface words appear in Words to Know.

MEET THE BOA CONSTRICTOR

How does a snake without **venom** catch its dinner? When night falls, boa constrictors slither over the ground and through the trees, looking for their next meal. They're great swimmers, but these heavy snakes like to spend most of their time on dry land.

BOA CONSTRICTOR

Boa constrictors live in Central and South America. They prefer **rain forests**, but they can be found in many different **habitats**. Boa constrictors also live in deserts and open savannas. They have special markings on their body that help them blend into their home.

BOA CONSTRICTORS CAN BE BROWN, RED, GREEN, YELLOW, OR BLACK.

MADE FOR HUNTING

Female boa constrictors are usually larger than males. Most boa constrictors grow to be about 6.5 to 10 feet (2 m to 3 m) long and weigh around 60 pounds (27 kg). Some larger boa constrictors have grown to be 13 feet (4 m) long and weighed around 100 pounds (45 kg)!

BOA CONSTRICTORS ARE IN THE SAME FAMILY AS THE ANACONDA.

When a boa constrictor sticks out its tongue, it's smelling the air! A boa constrictor's tongue pulls smells into its mouth. Inside the boa's mouth, there is an **organ** that helps it figure out what animals are nearby. This helps the boa constrictor find **prey** in the dark.

BOA CONSTRICTORS HAVE EXCELLENT EYESIGHT.

WHAT'S FOR DINNER?

Boa constrictors will eat anything they can fit in their mouth. Smaller boa constrictors will eat rats, birds, or squirrels. Larger ones can eat monkeys or wild pigs! Boa constrictors hide and wait for their prey, grabbing it in their teeth when it comes too close!

BOA CONSTRICTORS HAVE ROWS OF LONG, CURVED TEETH TO HELP THEM HOLD PREY.

A TIGHT SQUEEZE

After a boa constrictor's prey is caught, the boa gets ready to **squeeze**! A boa constrictor kills its prey by wrapping its long body around it and squeezing. This is how boa constrictors got their name. "Constrict" is another word for "squeeze."

IT CAN TAKE A BOA CONSTRICTOR UP TO SIX DAYS TO DIGEST A MEAL.

As soon as its prey is dead, the boa constrictor swallows it whole! Boa constrictors don't need to chew their food. Instead, their upper and lower jaws separate, letting them open their mouth extra wide. Then, they use their bottom jaw to "walk" the prey into their mouth!

IF A BOA CONSTRICTOR IS IN DANGER, IT WILL BITE TO KEEP ITSELF SAFE.

BABY BOAS

Boa constrictors don't lay eggs. Instead, female boa constrictors give birth to as many as 60 babies at one time! Newborn boas are around 24 inches (60 cm) long and are ready to hunt right away.

BOA CONSTRICTORS LIVE FOR AROUND 20 TO 30 YEARS.

BOA CONSTRICTORS AND PEOPLE

Boa constrictors are helpful hunters! In South America, some farmers use boa constrictors in their barns to hunt mice and rats. They're also kept as pets. Boa constrictors don't mind being around people and don't commonly attack them.

WHERE DO BOA CONSTRICTORS LIVE?

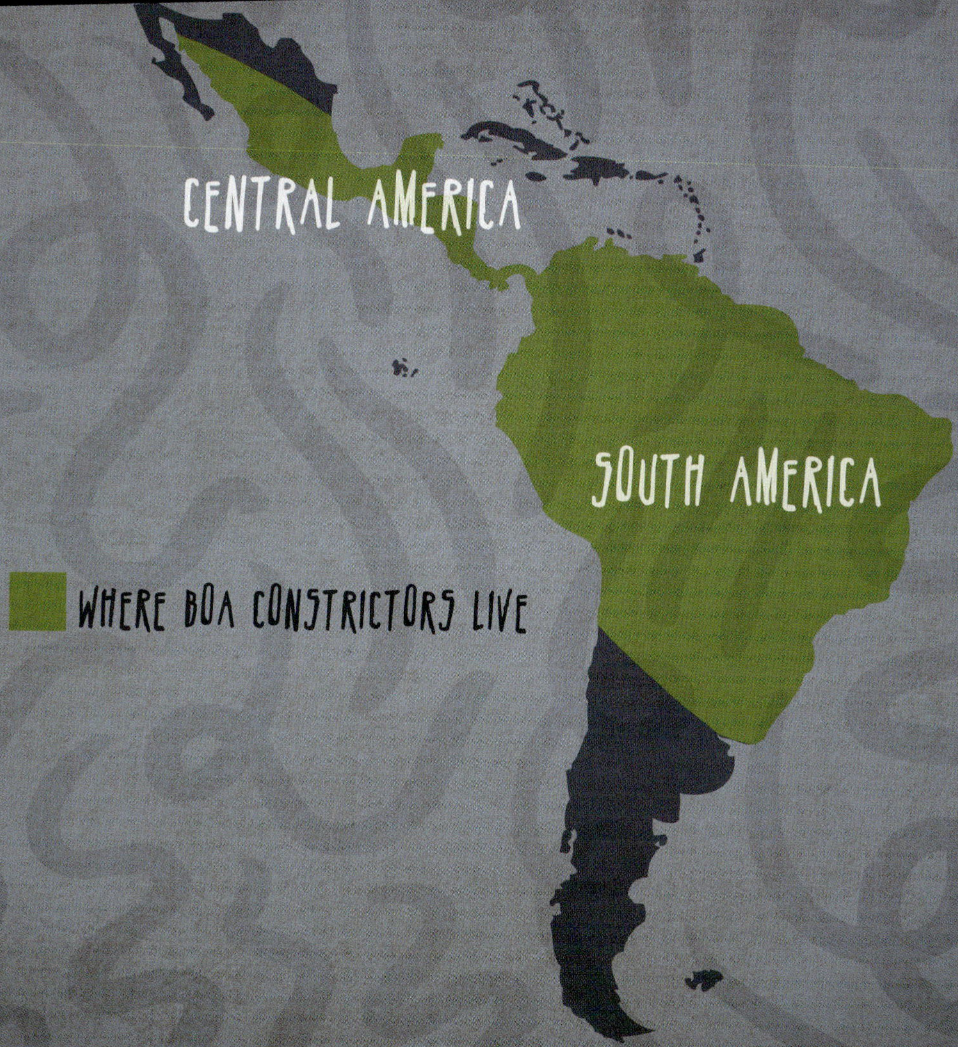

WORDS TO KNOW

digest To change eaten food into simpler forms that can be used by the body.

habitat The place or type of place where an animal naturally lives.

organ A part of the body that has a particular function.

prey An animal that is hunted or killed by another animal for food.

rain forest A forest that receives a lot of rain and has very tall trees.

squeeze To press together tightly.

venom Matter an animal makes in its body that can harm other animals.

FOR MORE INFORMATION

BOOKS

Hirsch, Rebecca E. *Boa Constrictors: Prey-Crushing Reptiles*. Minneapolis, MN: LernerClassroom, 2015.

Kenan, Tessa. *It's a Boa Constrictor!* Minneapolis, MN: Lerner Publications, 2017.

WEBSITES

National Geographic Kids
kids.nationalgeographic.com/animals/reptiles/boa-constrictor/
Watch videos and see pictures of boa constrictors in the wild.

San Diego Zoo
animals.sandiegozoo.org/animals/boa
Learn how to tell the difference between boas, pythons, and anacondas!

Publisher's note to educators and parents: Our editors have carefully reviewed these websites to ensure that they are suitable for students. Many websites change frequently, however, and we cannot guarantee that a site's future contents will continue to meet our high standards of quality and educational value. Be advised that students should be closely supervised whenever they access the internet.

INDEX